YOUR KNOWLEDGE HAS VALUE

- We will publish your bachelor's and master's thesis, essays and papers

- Your own eBook and book - sold worldwide in all relevant shops

- Earn money with each sale

Upload your text at www.GRIN.com and publish for free

Bibliographic information published by the German National Library:

The German National Library lists this publication in the National Bibliography;
detailed bibliographic data are available on the Internet at http://dnb.dnb.de .

Imprint:

Copyright © 2017 GRIN Verlag
Print and binding: Books on Demand GmbH, Norderstedt Germany
ISBN: 9783668900813

This book at GRIN:

https://www.grin.com/document/453908

Rony Antony

The Synthetic Antioxidant Ethoxyquin and its Effect on Fish Products

A Project Report in Advanced Food Chemistry

GRIN Verlag

Ethoxyquin and fish products

Project report in Advanced Food Chemistry

Rony Antony

Department of Food Science

Content

1.Abstract

The synthetic antioxidant Ethoxyquin (6-ethoxy-1,2-dihydro-2,2,4-trimethylquinoline) is widely used, specifically in the fishmeal industry to prevent the fishmeal from lipid oxidation. New implemented regulations of EU do not allow to incorporate any traces of EQ on feed in some countries within the Europe. But in many countries, it is used in some spices like rosemary and chilli to preserve their color. The toxicity of EQ is the major problem which creates some harmful effects on human and animal when animal are fed with the feed containing EQ. At the same time human are affected indirectly by other means. Here we tried to present the physical and chemical properties of EQ, its working mechanism as an antioxidant in fishmeal, positive and negative attributes of EQ on feed, feed ingredients, human and animals, regulations and requirements of incorporation of EQ on different feed, reauthorization part, replacement of EQ by some alternative natural and/or synthetic compounds as well.

2.Introduction

Three major synthetic antioxidants are used in food industry especially in processed feed such as fishmeal. These are Ethoxyquin (6-ethoxy-1,2-dihydro-2,2,4-trimethylquinoline) (EQ), Butylated Hydroxytoluene (2,6-di-tert-butyl-p-cresol) (BHT) and Butylated Hydroxyanisole (tert-butyl-4-hydroxyanisole) (BHA). Other synthetic antioxidants such as octylgallate and propylgallate have no significant role, even though among all the synthetic antioxidants, ethoxyquin is mostly used because of its low production costs and higher antioxidative effect (Alina Blaszoczyk *et al.* 2013). Ethoxyquin was initially developed in the rubber industry to prevent cracking of rubber while manufacturing due to the formation of isoprene (A.J.de Koning, 2002).

Autoxidation of lipids can occur spontaneously in a reaction with oxygen from the surrounding atmosphere oxidizing a lipid creating an alkyl radical as seen in the first equation in figure 2. The following two steps lead to a chain reaction where more alkyl radicals are formed.

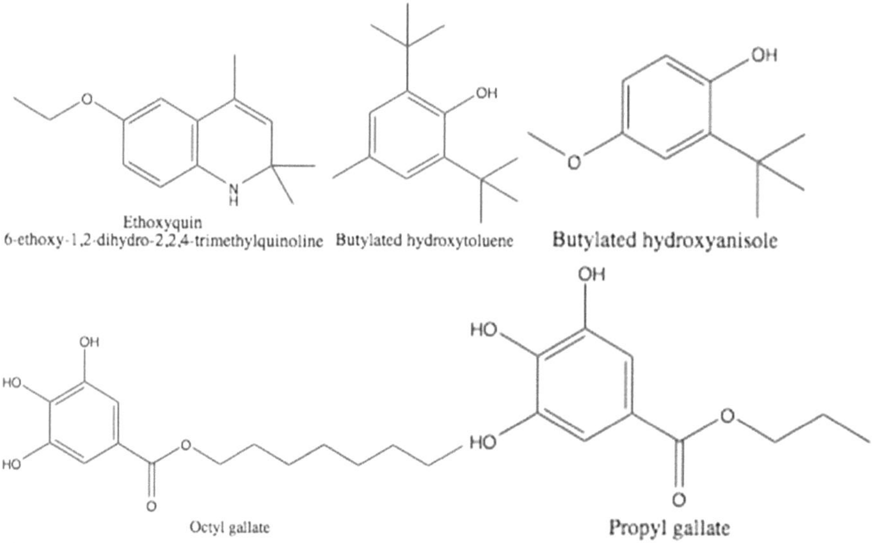

Figure 1. Synthetic antioxidants used to prevent oxidation in fishmeal

$$\text{Initiation: } RH \rightarrow R^{\bullet} + H^{\bullet} \qquad (4.1)$$

$$\text{Propagation: } R^{\bullet} + O_2 \rightarrow ROO^{\bullet} \qquad (4.2)$$

$$ROO^{\bullet} + RH \rightarrow ROOH + R^{\bullet} \qquad (4.3)$$

$$\text{Branching: } ROOH \rightarrow RO^{\bullet} + HO^{\bullet} \text{ Monomolecular} \qquad (4.4)$$

$$2\ ROOH \rightarrow ROO^{\bullet} + RO^{\bullet} + H_2O \text{ Bimolecular} \qquad (4.5)$$

$$\text{Termination: } ROO^{\bullet} + ROO^{\bullet} \rightarrow ROOR + O_2 \qquad (4.6)$$

$$R^{\bullet} + ROO^{\bullet} \rightarrow ROOR \qquad (4.7)$$

$$R^{\bullet} + RO^{\bullet} \rightarrow ROR \qquad (4.8)$$

$$R^{\bullet} + H^{\bullet} \rightarrow RH \qquad (4.9)$$

$$RO^{\bullet} + H^{\bullet} \rightarrow ROH \qquad (4.10)$$

$$ROO^{\bullet} + H^{\bullet} \rightarrow ROOH \qquad (4.11)$$

$$R^{\bullet} + HO^{\bullet} \rightarrow ROH \qquad (4.12)$$

Figure 2. Mechanisms of lipid autoxidation. (Márquez-Ruiz *et al.* 2012)

The longer the carbon-chain in poly unsaturated fatty acids (PUFAs) and if the double bonds are in a cis-configuration rather than trans-configuration the more prone they are to oxidation. (Márquez-Ruiz *et al.* 2012)

PUFAs are highly unsaturated hence they contain a lot of bis-allylic C-H bonds. Because of the double bonds on the carbon chains these bonds are weaker than allylic C-H bonds in saturated fatty acids. The bond dissociation energy (BDE) is 65 kcal/mol in PUFAs compared to 100 kcal/mol in saturated fatty acids.

Fish oil is highly unsaturated hence it has a tendency to undergo autoxidation. Autoxidation decreases energy- and nutritional value is decreased. So the stabilization of long chain polyunsaturated fatty acids (PUFA) (contains Omega-3 (Eicosapentaenoic acid) and Omega-6 (Docosahexaenoic acid) with ethoxyquin helps to maintain all the fatty acid characteristics of the fishmeal in a transportation and storage point of view. There is a probability to get overheated and spontaneously combust when fishmeal is stored due to heat produced during lipid autoxidation. (Olcott *et. al.*, 1962)). Stabilization of fishmeal is an integral part to avoid spontaneous combustion all through overseas transport and storage by the addition of between 400 and 1000 mg kg⁻¹ ethoxyquin, or of between 1000 and 4000 mg kg⁻¹ butylated hydroxytoluene at the period of production according to the International Maritime Organisation (IMO). This application should take place not more than 12 months prior to shipment, and the antioxidant application in fishmeal must be at least 100 mg kg⁻¹ at the time of shipment to preclude explosions (IMO 2003).

The autoxidation rate is also affected by the preprocessing history of the meal. For instance, when a poor quality raw material is used, then a faster oxidation can be seen. In many food systems, it is found that heme and nonheme iron promote or catalyse autoxidation. (Rhee, 1978, Toyoda *et al.*, 1982) Iron containing fish processing equipment are contacted fish meat part enhances lipid oxidation. (Lee *et al.*, 1977, Silberstein *et al.*, 1978). Ethoxyquin is the most common antioxidant used in fishmeal which is more efficient to prevent lipid oxidation and methionine oxidation rather than other antioxidants. (Gulbrandsen *et al.*, 1983).

3.Chemical structure

The molecular structure of ethoxyquin is a quinoline which is having a 1,2-dihydroquinoline bearing three methyl substituents at location 2, 2 and 4. In addition to that there is an ethoxy substituent at location 6. (6-ethoxy-1,2-dihydro-2,2,4-trimethylquinoline) The molecule is polar and the boiling and melting points are 123-125 °C and 0°C respectively (Blaszoczyk *et al.* 2013).

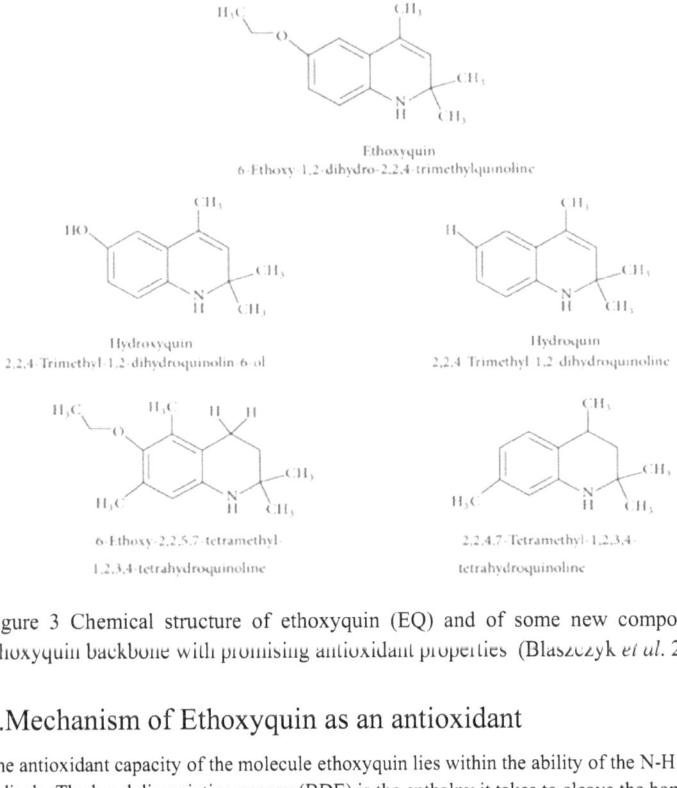

Ethoxyquin
6-Ethoxy-1,2-dihydro-2,2,4-trimethylquinoline

Hydroxyquin
2,2,4-Trimethyl-1,2-dihydroquinolin-6-ol

Hydroquin
2,2,4-Trimethyl-1,2-dihydroquinoline

6-Ethoxy-2,2,5,7-tetramethyl-1,2,3,4-tetrahydroquinoline

2,2,4,7-Tetramethyl-1,2,3,4-tetrahydroquinoline

Figure 3 Chemical structure of ethoxyquin (EQ) and of some new compounds synthesized on ethoxyquin backbone with promising antioxidant properties (Blaszczyk *et al.* 2013).

4.Mechanism of Ethoxyquin as an antioxidant

The antioxidant capacity of the molecule ethoxyquin lies within the ability of the N-H group to scavenge radicals. The bond dissociation energy (BDE) is the enthalpy it takes to cleave the homolytic bonds (Berton-Carabin 2014). The higher the value of BDE, the better the ability to scavenge radicals. The BDE of the N-H bond in EQ is 84 kcal mol^{-1} in water and 78 kcal mol^{-1} in gas phase. (Najafi *et al.* 2013)

The arylamines and secondary alkyl groups in ethoxyquin has antioxidant capacities but furthermore, it forms an oxidation product, an aminyl radical, which also have antioxidative properties when oxidized. (Scott, 1985), (Berger *et al.*, 1983), (Adamic *et al.*, 1969), (Adamic *et al.*, 1970),(Olcott *et al.*, 1970), Brownlie *et al.*, 1967). A simplified reaction mechanism of the formation of the antioxidative products of EQ-oxidation can be seen in figure 4. These oxidation products lead to the electron accepting chain-breaking antioxidants.

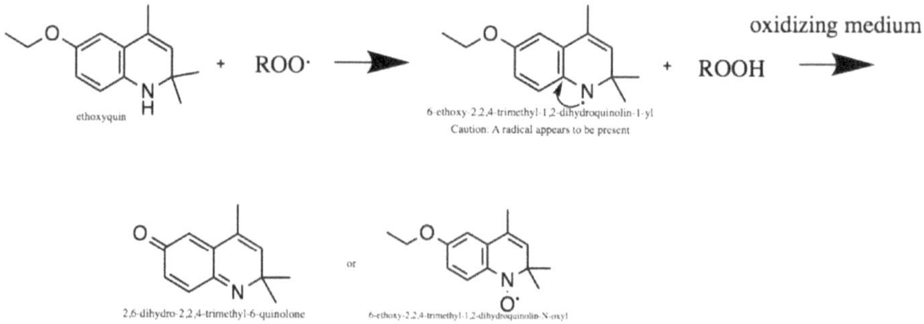

Figure 4. A schematic drawing of the radical scavenging by ethoxyquin and the products that also have antioxidative properties. Made from the reactions described by Blaszczyk *et al.* (2013)

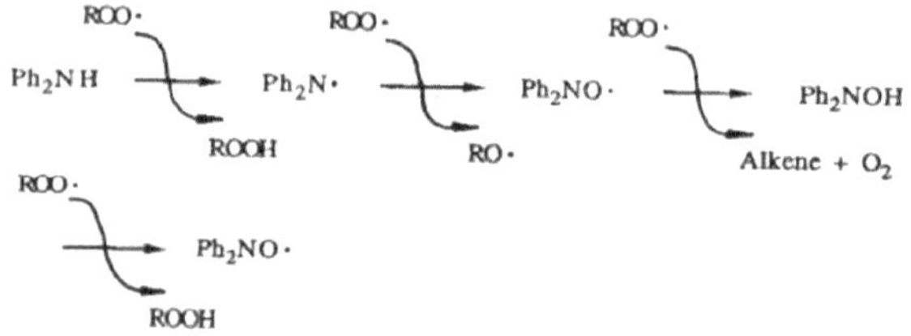

Figure 5. An illustration of a possible inhibition mechanism by a secondary aryl-amine (Berger *et al.*, 1983), (Denisov *et al.*, 1980)

Besides this reaction mechanism, a radical addition on the aromatic amine radical is also possible. The product of this reaction is the quinone-imine compound.(Adamic *et al.*, 1969). Formation of quinone-imine compound is shown in Figure 3).

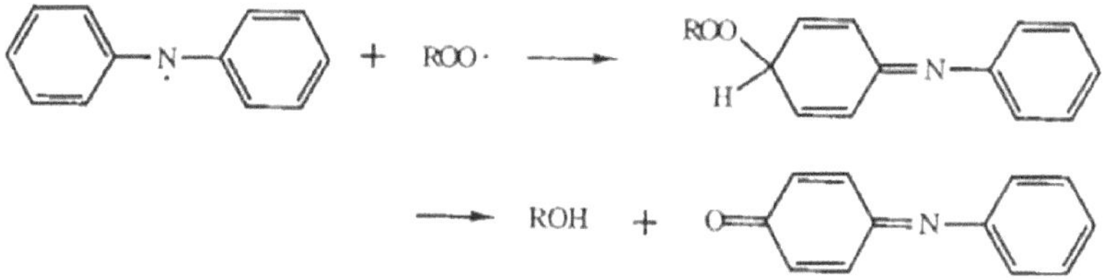

Figure 6. Quinone-imine formation (Adamic et al., 1969)

It has been observed that after oxidation of phenolic compounds, there are some oxidation products, that have stronger oxidation effects. These compounds are formed by the abstraction of hydrogen from the phenol or from different quinones like vitamin K, tocopherylquinone, rosmaryquinone or benzoquinone as shown in Figure 6. (Scott *et al.*, 1985), (Lindsey *et al.*, 1985), (Houlihan *et al.*, 1985).

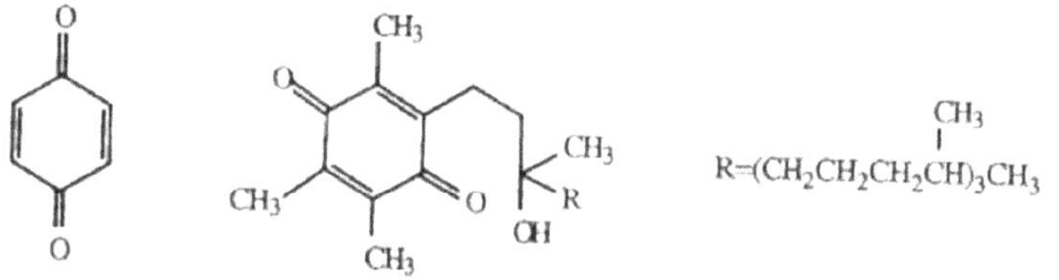

Benzoquinone   α-Tocopheroquinone

Figure 7. Different Quinone compounds, those are responsible for oxidation and producing oxidative compounds(Scott *et al.*, 1985), (Lindsey *et al.*, 1985), (Houlihan *et al.*, 1985).

5.Toxicity

Table 1: Effects of ethoxyquin detected after its oral intake in various animals or in humans (Contact exposure) (Alina Blaszczyk et al. 2013).

The harmful effect	Animals
Loss of weight	Marmosets[1], rats[2], dogs[2], mice[2], rabbits[3]
Changes in liver	Marmosets[1], rats[1,4], dogs[2], mice[2], broiler chickens[2], tilapia[5]
Changes in kidney	Marmosets[1], rats[2,6,7], dogs[2], broiler chickens[2,8]
Changes in alimentary duct	Marmosets[1], dogs[2], mice[2], broiler chickens[2,8]
Changes in urinary bladder	Rats[2]
Anemia	Marmosets[1]
Changes in mitochondria	Rats[2]
Lethargy	Rabbits[3]
Colored urine, skin, or fur	Dogs[2], rats[2]
Increase in mortality	Broiler chickens[2,8]
Effect on immunity	Tilapia[5]
Condition factor: the final body weight in relation to body length of fish	Large yellow croaker[9]
Allergy (contact exposure)	Humans[10,11,12]

The adverse effects of ethoxyquin was first reported in 1988, where dogs are the most susceptible to the detrimental effects of EQ which was reported by FDA Observed symptoms were dysfunction of the kidney, liver, reproductive system and the thyroid gland system (Dzanis D. A. 1991). Major symptoms like kidney and liver damage, alimentary duct alterations and weight loss are observed when animals are treated with a dosis of EQ above the maximum permitted concentration (Table 1). Reyes et al. (Reyes J. L. et al., 1995) and Hernandez et al. (Hernandez M. E. et al., 1993) point out that EQ interaction on biosystem is dose dependent, they found inhibition of renal sodium, potassium ions and ATPase activity involved in ion transport when they conducted experiment in bovine kidney and heart. (Hernandez M. E. et al., 1993). Experiments conducted in different strains of *Salmonella typhimurium* to assess the mutagenicity by using Ames test gave some positive results. (Rannug A. et al., 1984) & (Reddy B. S. et al., 1983). It was observed in animals exposed to EQ that i might promote or inhibit carcinogenic activity of known carcinogens (Alina Blaszczyk et al. 2013). Programmed cell death called apoptosis or necrosis by the effects of EQ may further damage genetic material such as DNA and chromosomes (Alina Blaszczyk et al. 2013). Apoptosis in lymphocytes while in-vitro culture was observed by Blaszczyk (Blaszczyk et al., 2005). Overall, serious biological consequences such as chromosome aberrations were observed in Chinese hamster ovary cells and human lymphocytes (Rabbitts T. H., 1994).

Some case studies reveal that the use of EQ on animal feed has both direct and indirect impact on human on a larger scale. Van Hecke (Van Hecke, 1977) and Savini *et al.*, (1989) studied that people handling EQ had an itchy, scaly, erythematous dermatitis with irregular oozing eruptions. Zachariae (1978) found that people handling EQ was suffering from exfoliative dermatitis after ten days of EQ exposure. (Brandao 1983), (Burrows 1975) and (Wood and Fulton 1975) point out that some workers handling feed material with EQ, began suffering from chronic and acute dermatitis which spread throughout their body parts.

Table 2 Effects of ethoxyquin intake of rats (Arthur. D. 1990)

Ethoxyquin Induction Tumor Formation in F344 Rats

SITE	CARCINOGENIC AGENT	PERCENT ETHOXYQUIN IN DIET (DAYS)	INCREASE TUMOR FORMATION (YES/NO)	REFERENCE
Forestomach	BNA	0.25% (7)	YES	Hirose, et al., 1986
Urinary Bladder	BNN	0.5-0.125% (154)	NO	Fukishima, et al., 1987
Urinary Bladder	BNN	0.8% (224)	YES	Fukishima, et al., 1987
Urinary Bladder	BNN	0.8% (22)	YES	Miyata, et al., 1985
Urinary Bladder	BNN	0.8% (203)	YES	Ito, et al., 1986
Glandular Stomach	MNNG	1.0% (56)	YES	Takahashi, et al., 1986
Esophageal	DBN	0.8% (252)	YES	Fukishima, et al., 1987
Distal Colon	DMH	0.8% (252)	NO	Ito, et al., 1986
Kidney	EHEN	0.8% (203)	YES	Ito, et al., 1986

6.Regulations and requirements

According to International Maritime Organisation (IMO), stabilization of fishmeal is an integral part to avoid spontaneous combustion all through overseas transport and storage by the addition of between 400 and 1000 mg kg^{-1} ethoxyquin, or of between 1000 and 4000 mg kg^{-1} butylated hydroxytoluene since ethoxyquin has the ability to stabilize the long chain polyunsaturated fatty acids in the fishmeal to prevent them from oxidizing. The stabilization process was initiated in 1970s. Almost 66% of world's fishmeal production was preserved using ethoxyquin.

In Code of Federal Regulations (2003) the legislation of antioxidants in fishmeal is written and it dictates certain minimal limits of ethoxyquin in different cases; at production the treatment should be at least 400 ppm. If the fishmeal contains more than 12% fat the minimal treatment limit is 1000 ppm and at shipment 100 ppm of EQ is required to be present in the product.

According to IFFO (2016) EQ is in a reauthorization process because maximum limits in compound feed has been altered as the proportion of fishmeal in compound feed has been reduced.

Currently compound meal has a very low content of compound feed. Hence the average concentration level of ethoxyquin for the fish feed in Europe has been significantly reduced even though the fish feed has a higher maximum limit. Most of the producers of fish feed are supportive towards the ethoxyquin reauthorization process. For that they are strongly motivated to reduce the content of ethoxyquin in compound fish feed. In 2003 the reauthorization process was introduced to evaluate all food additives considering their safety aspects. The pertinent regulation is described in

Directive (EC) No 1831/2003 which state that the usage of complete animal feed additives should be reviewed, and approved endorsements should be time-limited to allow regular reviews. After considering humans and animals as well as environmental health, any action should be on the basis of preventative or precautionary principle.

The European Food Safety Authority (EFSA) (2015) start the safety data review as a basis for the reauthorization process. Authorization of ethoxyquin for the use in animal feed is under Directive 70/524 legislation, it was authorized that ethoxyquin shall be used in animal feed, but with a limit of 150 mg/kg. A reauthorization request was submitted in 2010.

As part of the reauthorization process a debate needs to be conducted. For that the European Commission acts as standing committee (SCoPAFF- Plants, Animals, Food and Feed) and Animal Nutrition Section to debate the authorization for ethoxyquin application. Within a period of 3 months of receival at the EFSA Opinion, the European Commission will prepare a draft regulation in which commission may grant or deny authorization. There is a provision for the extension of deadline in complex cases. There is a monthly meeting conducted by SCoPAFF Animal Nutrition section and there is discussion furtherance on the safety and economic significance of ethoxyquin in March. At this point, it is possible to elucidate further if European commission is still on the way making a better judgment for the authorization. (European Commission 2016).

But in November 2015, it was found that the review of European Food Safety Authority (EFSA) about ethoxyquin usage in animal feed was not conclusive due to several knowledge gaps and data deficits. EFSA has already published an opinion on ethoxyquin reauthorization. Still the opinion of EFSA has to be considered by the European committee whether the use of ethoxyquin should be authorized or not. Political pressure is one of the reasons why the European Commission will take EFSA's side to provide sufficient grounds to sanction authorization of ethoxyquin on precautionary grounds. Andersen (2017b)

The effect of withdrawal of ethoxyquin has to be studied well so IFFO (The Marine Ingredients Organization), wants to conduct a study enlightening the consequences of ethoxyquin withdrawal from feed, fishmeal and aquaculture industries. The study will give a solid background of the effect of ethoxyquin on various feedstuff by providing adequate data which can re-fill the gaps and finalize the decision.

IFFO has a collaboration between European Trade Bodies FEFANA and FEFAC, abbreviations stand for feed ingredients and feed consecutively, liaise with the European Commission throughout the time period. IFFO has been on the way to look at different scope for minimizing the use of ethoxyquin in fishmeal without affecting quality, safety or any other parameters. The outcome of reauthorization may be a reduced incorporation of ethoxyquin in feed or feed ingredients.

In order to study accelerated fishmeal stabilities, IFFO is also experimenting with synthetic antioxidants (BHT-Butylated Hydroxytoluene) along with natural herbs (Rosemary extract and Tocopherol) with antioxidative properties, to get an efficient product of antioxidant, which it is possible to incorporate in feedstuff. Rosemary extract and tocopherol act as stabilizing agents. They conduct different storage trials for estimating the efficiency of combined formulation in fishmeal.

A collaboration between IMO and United Nations Transport for Dangerous Good Committee (UN-TDG) to amend present legislation dealing with inclusion levels of antioxidants ethoxyquin and

Butylated Hydroxytoluene in fishmeal, are on the way to include alternative antioxidants. The process of discussion is continuing.

However EU authorities may prohibit or suspend the ethoxyquin incorporation approval to feed instantaneously. SCoPAFF conducts meeting to take a decision with regards to this. If European Authorities tend to suspend the approval of ethoxyquin, IFFO may recommend EU members to consider different substitutes to ethoxyquin and their potential impacts.

7.Limits of ethoxyquin residuals in human food

EQ level is controlled and regulated by FDA (Food and Drug Administration in 21 CFR parts 172.140, 177.2600, 573.380 and 573.400 and 40 CFR part 180.178 (Federal Register Office, 1988,1989, 1990). As per CFR provisions (FDA 2016) Ethoxyquin, it is quoted that,

(a) Ethoxyquin (1,2-dihydro-6-ethoxy-2,2,4-trimethylquinoline) may be safely used as an antioxidant for preservation of color in the production of chili powder, paprika, and ground chili at levels not in excess of 100 parts per million.

(b) In order to assure the safe use of the additive in feed prepared in accordance with 573.380 and 573.400 of this chapter, tolerances are established for residues of ethoxyquin in or on edible products of animals as follows:

Table 3 Limits of ethoxyquin in different food stuff in USA. (Code of Federal Regulations 2016)

1	5 (ppm) parts per million in or on the uncooked fat of meat from animals except poultry.
2	3 (ppm) parts per million in or on the uncooked liver and fat of poultry.
3	0.5 (ppm) part per million in or on the uncooked muscle meat of animals.
4	0.5 (ppm) part per million in poultry eggs.
5	Zero (0 ppm) in milk.

According to EC Council Directive (70/524/EEC), for European Countries (EC), maximum amount of ethoxyquin in animal feedstuffs is 150 mg/kg.

According to EC council directive 81/962/EEC, addition of EQ to foods as an antioxidant is not permitted. National food legislation in European Countries (EC) conducted a survey by the UK Food Research Association (Research Report No: 1993-7; Revised February 1995) concluded the situations in various countries as follows (Quoted):

Denmark:	EQ is not permitted for use in any foods for human consumption.
France:	EQ is a permitted agent for treating products after harvest and may be used for treating pears and apples, the maximum residue is 3 mg/kg in the whole fruit, from soaking in an EQ solution at 0.25-0.35%. It is not permitted for any other food use.
Italy:	EQ is not permitted for food use.
Netherlands:	No authorization is given for the use of EQ on fruit or other foods.
Spain:	EQ is not approved as an additive for foodstuffs.
Portugal:	EQ is not permitted as an additive for foodstuffs.
Germany:	EQ is not permitted as a direct food additive but is permitted as a pesticide with a maximum permitted residue level for all plant foodstuffs at 0.01 ppm (mg/kg).
Greece:	EQ cannot be used in human foodstuffs.
UK:	EQ is permitted for use as a pesticide on apples and pears to maximum 3 mg/kg.
Ireland:	EQ is permitted for use as a pesticide on apples and pears to maximum 3 mg/kg.
Belgium:	EQ is not permitted as an additive for human food.
Luxembourg:	EQ is not permitted as an additive for human food.
Norway:	EQ is not listed in the Positive Additives Lists and therefore would not be permitted for use in any foods for human consumption. Norway is currently bringing its animal feed legislation and other legislation to be equivalent to EC Directives.
Finland:	EQ is not listed in Finland's Positive Additives List and therefore would not be permitted for use in any foods for human consumption. EQ is permitted in animal feed to maximum 100 mg/kg.

Sweden:	EQ is not permitted for use on fruit, but is permitted at maximum 100 mg/kg in capsicum-based spice blends and spice extracts. For animal feeds, a maximum level of 100 mg/kg applies, singly or together with other permitted antioxidants. EQ is not to be used in combination with formic acid. (EQ and formic acid can stimulate the production of aflatoxin in green feed - see *Foudin et al, 1978*).
USA:	EQ is a permitted antioxidant for the preservation of colour in the production of chilli powder, paprika and ground chilli at levels not in excess of 100 mg/kg.

EQ is permitted in animal feedingstuffs to a maximum level of 150 mg/kg. In order to provide for the safe use of the additive in animal feeding stuffs tolerances are established for residues of EQ in or on edible products of animals as follows:

> 5 ppm in or on the uncooked fat of meat from animals except poultry.
>
> 3 ppm in or on the uncooked liver and fat of poultry.
>
> 0.5 ppm in or on the uncooked muscle meat of animals.
>
> 0.5 ppm in poultry eggs.
>
> 0 ppm in milk.

Mexico:	EQ is not listed as a permitted additive for use in foodstuffs.
Chile:	EQ is not listed as a permitted additive for use in foodstuffs.
Japan:	The information service of the Leatherhead Food Research Association was not aware of any regulations governing the use of EQ in foods or animal feeds, the authorities in Japan need to be contacted for advice on the acceptability of its use.

Based on the office of the Federal Register, 1990, the maximum permitted quantity of EQ in treated poultry feed and forage crops is 150 ppm. An acceptable daily intake of EQ for human is 0.005 mg/kg (WHO, 2012). Canada, UK and New Zealand have established a level of tolerance of 3 ppm for EQ on apples. Based on Canadian tolerance level, 3 ppm of EQ on apples, poultry liver and 0.5 ppm of EQ in poultry, eggs and meat is also established (FAO/WHO, 1970).

8.Natural alternatives

Ethoxyquin can cause a negative health effects on the animals that feed on the fishmeal treated with the antioxidant. (Lahabanjong *et al.* 2009) Because of increased knowledge about problems in using ethoxyquin as an antioxidant, natural alternatives such as vitamin E, vitamin C, peptides, amino acids, chitooligosaccharides (COS), astaxanthin, carotenoids, sulfated polysaccharides (SP), phlorotannins, phenolics, flavones and isoflavones are being evaluated (Aklakur 2016).

Natural antioxidants can be derived from plants where the chemical object of the compounds is to protect the organism from oxidative stress. (Brewer 2011)

Plant derived antioxidants include four major groups: phenolic acids which trap free radicals, phenolic diterpenes, flavonoids, that scavenge free radicals and chelate metals and volatile oils. Their strong antioxidative properties lie within the ability to stop the oxidation chain reaction by binding and transforming the free radicals and the effectiveness is somehow proportional to the number of –OH groups on the compounds. Because of the high number of aromatic rings in the natural antioxidants they often contain a high number of –OH groups which make them good alternatives to synthetic compounds.

The antioxidative property of antioxidants can be measured using various methods including measurement of oxygen consumption in emulsions, oxygen radical absorbance capacity (ORAC) values, iodine value, anisidine value and measurement of absorbance at 234 and 268 nm indicating early oxidation as conjugated dienes and trienes absorb light at these wavelengths (Brewer 2011).

According to Aklakur (2016), many natural alternatives could be used successfully in the fishmeal industry but because of a high production costs due to limited raw material supply and no standardized production, the antioxidants are still simply too costly to introduce to the industry.

9.Discussion

A vast number of antioxidants, are used in the food industry. Those antioxidants are either natural or synthetically produced. Synthetic antioxidant like ethoxyquin has been used for more than fifty years, because of its low production cost and high antioxidant effect. It can prevent oxidation of lipids and many vitamins. According to EC Council Directive (70/524/EEC), for European Countries (EC), the maximum amount of ethoxyquin in animal feedstuffs is 150 mg/kg. The antioxidant property of ethoxyquin is highly used in the production of fishmeal to stabilize the fatty acids. Two different types of antioxidant mechanisms are involved mainly in preventing lipid oxidation of fat by ethoxyquin. These are a radical scavenging mechanism and a radical addition mechanism. The oxidation products produced have great antioxidant capacities, which will aid to protect fishmeal subsequently during transportation. One of the problems faced by fishmeal industry is the oxidation of fishmeal during transportation and storage which affects the quality and nutritive value of fishmeal tremendously. Oxidation of fishmeal is possible to result in spontaneous combustion because of high heat production due to lipid oxidation of fishmeal. Hence the application of antioxidants is of high importance in these products.

However, ethoxyquin has can have a negative impact when it is used as a feed additive in foodstuffs. Carcinogenic and immunotoxic and mutagenic effect of EQ is studied well and researches are on the way to assess the toxigenicity of EQ in a greater extent. EQ is especially used in pet foods. Compared to other animals, dogs are more prone to the toxic effect of EQ and researchers found a number of carcinogenic compounds and other compounds toxic to internal organs of dogs. With regards to humans, individuals dealing with this antioxidant in feed formulation processes also experience problems like skin irritations and acute dermatitis. However, the effect of EQ on human is not highly explored and further research is essential to know the exact behaviour of this antioxidant on the human organism because humans are exposed to this antioxidant indirectly through the fish fed with EQ-treated feedstuff.

Another challenge is the detection of concencentration of EQ in feedstuffs, since food is a very complex system and so many parameters have to be considered well. Different chemical analytical methods such as oxygen radical absorbance capacity (ORAC) values, iodine value, anisidine value are extensively used for the EQ level detection. The product and byproducts formed by the interaction may influence the antioxidant capacity (effectiveness and efficiencies) of EQ. Some byproducts or main product may or may not promote the antioxidant mechanism. At the same time, ambient storage temperature of fishmeal and other physical as well as chemical factors has to be considered well.

According to International Maritime Organization (IMO), stabilization of fishmeal with antioxidant is highly essential. The Marine Ingredients Organization (IFFO) is in a reauthorization process since they consider different relevant parameters such as humans and animals as well as environmental health are concerned. The disadvantages of EQ outweigh the advantages when the toxic effect of EQ on human and animals is scrutinized.

Because of the negative health effects of EQ, methods of reducing the use of this synthetic antioxidant are being investigated. To exclusively use natural antioxidants is currently not a pragmatic approach since most of the natural antioxidants have a relatively short antioxidant effect on the fishmeal, besides it is very costly hence not economically feasible. One solution is to minimize the concentration of synthetic antioxidant like EQ by using some natural alternatives. Another possibility is to develop a new chemical compound with less toxicity and high antioxidant capacity which should be specified for fishmeal. A compound corresponding to 'Naturone' which is widely used in the poultry industry. In order to study accelerated fishmeal stabilities, IFFO is also experimenting with synthetic antioxidants (BHT-Butylated Hydroxytoluene) along with natural herbs (Rosemary extract and Tocopherol) with antioxidative properties, to get an efficient product of antioxidant, which it is possible to incorporate in feedstuff. Rosemary extract and tocopherol act as stabilizing agents.

10.Conclusion

Ethoxyquin has been widely used in fishmeal as an antioxidant and its antioxidant capacity through free radical scavenging mechanism has been widely understood. However when toxigenicity in human and animals is considered, the use of ethoxyquin is still in a bewildered stage for a decision. A combination of multiple antioxidants (either natural or synthetic or a combination of natural and synthetic) with less toxicity and stability along with some emulsifying or chelating agents may provide some good results. For that extensive data collection and research have to be conducted. Through this process, stabilization and reauthorization is possible to a greater extent and the present huddle can be minimized. Knowing the toxicity, the regulating body shall consider the maximum inclusion level of EQ on feed and feed ingredients. When finding the maximum allowable or permissible level dangerous to the health of animals, the present maximum limit can be amended. Labelling of EQ incorporated feed is mandatory and provide a precautionary sign to the customers whether to accept or avoid these EQ incorporated feed for their pets.

11.References

Adamic, K., Bowman, D. F., & Ingold, K. U. (1970). The inhibition of autoxidation by aromatic amines. *Journal of the American Oil Chemists Society*, *47*(4), 109-111.

Adamic, K., Dunn, M. and Ingold (1969). Formation of diphenyl nitroxide in diphenylamine inhibited autoxidations, K.U. *Canadian Journal of Chemistry*, 47, 287-294.

Adamic, K.and Ingold, (1969). Formation of radicals in the amine inhibited decomposition of t-butyl hydroperoxide, *Canadian Journal of Chemistry*, 47, 295-299.

Andersen, M.L. (2017a). Slide show on Lip ox 1 2017 adv food chem.

Andersen, M.L. (2017b). Personal comment. 28. april 2017.

Arthur. D. 1990., Ethoxyquin, Executive Summary of Safety and Toxicity Information, National Toxicology Program, CAS No: 91-53-2.

Berger.H., Bolsman, T.A.B.M. and Brouwer, D.M., 1983. Developments in polymer stabilization, Scott, G. (ed), Vol. 6, Chap.1, App.Sci.Publ., London.

Berton-Carabin, C. C., Ropers, M. H., & Genot, C. (2014). Lipid Oxidation in Oil-in-Water Emulsions: Involvement of the Interfacial Layer. *Comprehensive Reviews in Food Science and Food Safety*, *13*(5), 945-977.

Brandão, F. M. (1983). Contact dermatitis to ethoxyquin. *Contact dermatitis*, *9*(3), 240-240.

Brewer, M. S. (2011). Natural antioxidants: sources, compounds, mechanisms of action, and potential applications. *Comprehensive Reviews In Food Science and Food Safety*, *10*(4), 221-247.

Brownlie, I.T. and Ingold, 1967.The inhibited autoxidation of styrene. Part VII. Inhibition by nitroxides and hydroxylamines, K.U. *Canadian Journal of Chemistry*, 45, 2427-2432.

Burrows, D. (1975). Contact dermatitis in animal feed mill workers. *British Journal of Dermatology*, *92*(2), 167-170.

Błaszczyk, A., Augustyniak, A., & Skolimowski, J. (2013). Ethoxyquin: an antioxidant used in animal feed. *International journal of food science*, *2013*.

Błaszczyk, A. (2006). DNA damage induced by ethoxyquin in human peripheral lymphocytes. *Toxicology letters*, *163*(1), 77-83.

Błaszczyk, A. L. I. N. A., & Skolimowski, J. (2005). Apoptosis and cytotoxicity caused by ethoxyquin and two of its salts. *Cellular and Molecular Biology Letters*, *10*(1), 15-21.

Code of Federal Regulations (2003), "Ethoxyquin and fishmeal, Title 46,Volume 5 (Revised as of October 1, 2003), Part 148, Subpart 148.04, Section 148.04-9.

Code of Federal Regulations (2016), Food Additives Permitted for Direct Addition to Food for Human Consumption, Title 21, Volume 3 (Revised as of April 1, 2016), 21CFR172, Part 172.

de Koning, A. J. (2002). The antioxidant ethoxyquin and its analogues: a review. *International Journal of Food Properties*, *5*(2), 451-461.

Denisov, E.T. 1980. Developments in Polymer Stabilisation, Scott, G. (ed), Vol. 3, Chap. 1, Appl. Sci. Publ., London.

Dzanis, D. A. (1991). Safety of ethoxyquin in dog foods. *The Journal of nutrition*, *121*(11 Suppl), S163-S164.

European Commission (2016). Standing Committee on Plants, Animals, Food and Feed Section Animal Nutrition, Health and Food Safety Directorate General, Ares (2016) 936544. Available at; http://ec.europa.eu/food/animals/docs/reg-com_ani-nutrit_20160308_agenda.pdf (Accessed June 2017; 9th).

European Food Safety Authority (EFSA) (2015). Ethoxyquin: EFSA safety assessment inconclusive, Available at; http://www.efsa.europa.eu/en/press/news/151118 3

FDA (2016). Food additives permitted for direct addition to food for human consumption. *Code of federal regulations. Food for human consumption*. Title 21, volume 3.

Hernandez, M. E., Reyes, J. L., Gomez-Lojero, C., Sayavedra, M. S., & Melendez, E. (1993). Inhibition of the renal uptake of p-aminohippurate and tetraethylammonium by the antioxidant ethoxyquin in the rat. *Food and chemical toxicology*, *31*(5), 363-367.

Houlihan, C.M., HO, C.T. and Chang, S.S., (1985).The structure of rosmariquinone — A new antioxidant isolated from *Rosmarinus officinalis L,Journal of American Oil Chemistry Society*, 62, 96-98.

IFFO (2016) Ethoxyquin and its use as antioxidant in fishmeal and fish feeds. *The marine ingredients organisation*. Notes for members. Available online at: http://www.iffo.net/system/files/Uses%20of%20ethoxyquin%20and%20alternatives%201993-7.pdf

International Fishmeal & Oil Manufacturers Association (IFOMA) 1995, Present and Future Uses of EQ and Alternative Antioxidants for the Stabilisation of fishmeal, Research Report No: 1993-7; October 1993.

International Maritime Organisation (IMO). 2003. Review of the BC Code, including evaluation of properties of solid bulk cargoes. Report of the Working Group at DSC 7. I:\DSC\8\4.doc. London (UK): IMO.

Laohabanjong, R., Tantikitti, C., Benjakul, S., Supamattaya, K., & Boonyaratpalin, M. (2009). Lipid oxidation in fishmeal stored under different conditions on growth, feed efficiency and hepatopancreatic cells of black tiger shrimp (*Penaeus monodon*). *Aquaculture, 286*(3), 283-289.

Lee, C.M. and Toledo, R.T, 1977. Degradation of Fish Muscle During Mechanical Deboning and Storage with Emphasis on Lipid Oxidation. *Journal of Food Science*, 42, 1646-1649.

Lindsey, J.A., Zhang, H., Kaseki, H., Morisaki, N., Sato, T. and Cornwell, D.G., 1985. *Lipids*, 20, 151-157.

Márquez-Ruiz, G., Holgado, F., & Velasco, J. (2013). Mechanisms of oxidation in food lipids. In *Food Oxidants and Antioxidants: Chemical, Biological, and Functional Properties* (pp. 79-114). CRC Press.

Najafi, M., Najafi, M., & Najafi, M. (2013). Predicting the substituent and solvent effects on the radical scavenger activity of ethoxyquin derivatives: a DFT/B3LYP study. *Canadian Journal of Chemistry, 91*(6), 457-464.

Olcott, 1962. H.S. Symposium on Foods: Lipids and their oxidation, Schultz, H.W., Day, E.A. and Sinnhuber, R.O. (eds), Chap. 9, Avi Publishing, Westport.

Olcott, H.S., Veen, J. van der and Koide, T. Bull., 1970. Role of Individual Phospholipids as Antioxidants. *Jpn. Soc. Sci. Fish.*, 36, 844-846.

Rabbitts T. H. (1994), Chromosomal translocations in human cancer, *Nature, vol. 372*, no. 6502, 143-149.

Rannug, A., Rannug, U., & Ramel, C. (1984). Genotoxic effects of additives in synthetic elastomers with special consideration to the mechanism of action of thiurams and dithiocarbamates. *Progress in clinical and biological research, 141*, 407.

Reddy, B. S., Hanson, D., Mathews, L., & Sharma, C. (1983). Effect of micronutrients, antioxidants and related compounds on the mutagenicity of 3, 2′-dimethyl-4-aminobiphenyl, a colon and breast carcinogen. *Food and Chemical Toxicology, 21*(2), 129-132.

Reyes J. L., Elisabeth Hernandez M., Melendez E., Gomez-Lojero C. (1995), Inhibitory effect of antioxidant EQ on electron transport in the mitochondrial respiratory chain, *Biochemical pharmacology,* vol. 49, no. 3. 283-289.

Rhee, 1978. K.S. Factors Affecting Oxygen Uptake in Model Systems Used for Investigating Lipid Peroxidation in Meat, *Journal of Food Science,* 43, 6-9.

Savini, C., Morelli, R., Piancastelli, E., & Restani, S. (1989). Contact dermatitis due to ethoxyquin. *Contact dermatitis, 21*(5), 342-343.

Silberstein, D.A. and Lillard, D.A., Factors Affecting the Autoxidation of Lipids in Mechanically Deboned Fish, *Journal of Food Science,* 43, 764-766.

Toyoda, I., Terao, J. and Matsushita, 1982. Hydroperoxides formed by ferrous ion-catalyzed oxidation of methyl linolenate. *S. Lipids,* 17, 84-90.

World Health Organization/Food and Agricultural Organization, (WHO/FAO) (1970), "Ethoxyquin." Evaluations of Some Pesticide Residues in Food, 103115.

World Health Organization (WHO) (2012), (Vettorazzi, G.) Inventory of Evaluations Performed By the Joint Meeting on Pesticide Residues. Available online at: http://apps.who.int/pesticide-residues-jmpr-database/pesticide?name=ETHOXYQUIN

Zachariae, H., (1978), "Ethoxyquin Dermatitis." *Contact Dermatitis,* Vol. 4, No. 2. 117-118.

Abbreviations

BHA	Butylated Hydroxyanisole
BHT	Butylated Hydroxytoluene
BBN	N-butyl-N-(4-hydroxylbutyl) nitrosamine
CAS	Chemical Abstracts Service
CFR	Code of Federal Regulation
COS	Chitooligosaccharides
CSP	European Food Safety Authority
DBN	N, N-dibutylnitrosamine
DMH	1,2 dimethylhydrazine
DNA	Deoxy Ribo Nucleic acid
EC	European Commision
EEC	European Economic Community
EFSA	European Food Safety Authority
EHEN	N-ethyl-Nhydroxyethylnitrosamine
EQ	Ethoxyquin
EU	European Union
FAO	Food and Agriculture Organization
FDA	Food and Drug Administration
FEFAC	European Trade Bodies For Feed
FEFANA	European Trade Bodies For Feed Ingredients
IFFO	The Marine Ingredients Organization
IMO	International Maritime Organization
MNNG	N-methyl-N-Nitro-N-nitrosoquanidine
-OH	Hydroxyl Group
ORAC	Oxygen radical absorbance capacity
PUFA	Polyunsaturated Fattyacids
SCoPAFF	Animal Nutrition section of the Standing Committee on Plant Animal Food and Feed
UK	United Kingdom
UN-TDG	United Nations Transport for Dangerous Good Committee
WHO	World Health Organization
ppm	Parts per million
mg	milligram
kg	kilogram
nm	nano meter